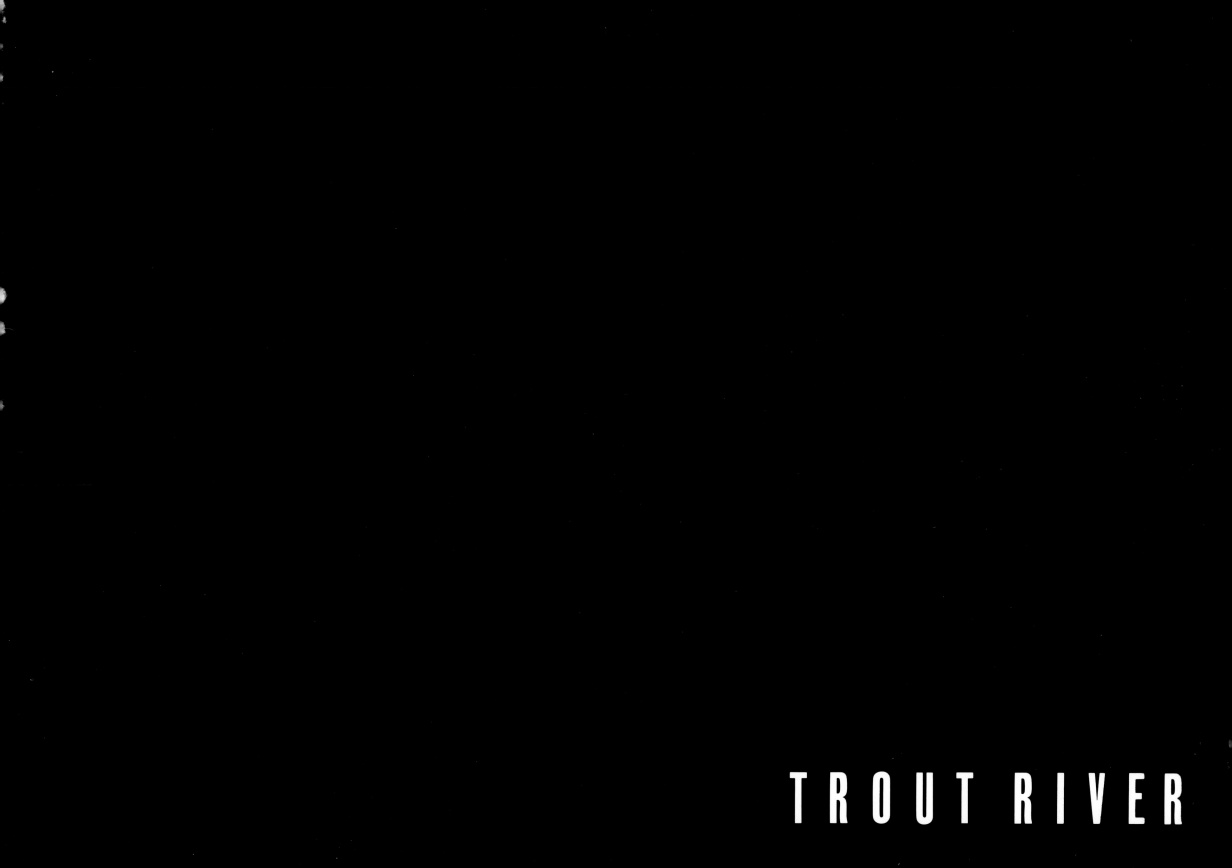

TROUT RIVER

HARRY N. ABRAMS, INC., PUBLISHERS, NEW YORK

TEXT BY NICK LYONS

CONCEIVED AND PHOTOGRAPHED
BY LARRY MADISON

TROUT RIVER

To love a river is to know it in all its moments, in all seasons. For a river is not one thing but many. It is alive and infinitely varied. It flows in a thousand configurations.

One river starts high on some desolate mountain slope in a miniature spring among the mosses; small streams, fed by springs, or snowmelt, or rainwater, rush to join it—and the little river dashes brightly down a rocky path to the valley below. Another gushes up full blown from a desert floor. Still another seeps up from a bog under a beaver pond and flows, sluggish and dark, harboring small, brilliantly colored brook trout beneath its overhanging banks.

A river may be the tailwater of a mighty reservoir, which diminishes or swells at the need of man, or the outflow from a lake fed by innumerable springs. Rivers are swift or slow, wide or narrow, sometimes straight, but chiefly meandering—as is their wont—and always with myriad variations. Rivers are bent by terrain and by season, and their changing character affects the world around them—soil and landscape, flora and fauna, even people.

There is no true place to stop or start a portrait
of the life of a river any more than there is a
beginning or end to a circle. It may be best
simply to begin when the year begins,
when the river and all around it are bound in
ice and snow.

Locked within the cold, like a bear in its cave,
the river seems to hibernate. Ice has
methodically built up in layers, in terrible
symmetries, upon the rocks, and the banks
steadily encroach upon the river's domain.
Storms have buffeted the river, redesigning its
path, surging its flow, setting it to cut a new
route through the forest and valley.

This is a severe time for the river and its denizens: those birds that have not left for warmer locales live a marginal, fugitive life; the groundhog and raccoon have hibernated; near the bottom of the deeper pools, the trout rest—slow and still. But the river has fought to remain free and alive. Beneath the snow-blanketed ice and in every open channel it can find, the water flows on, cold and dark.

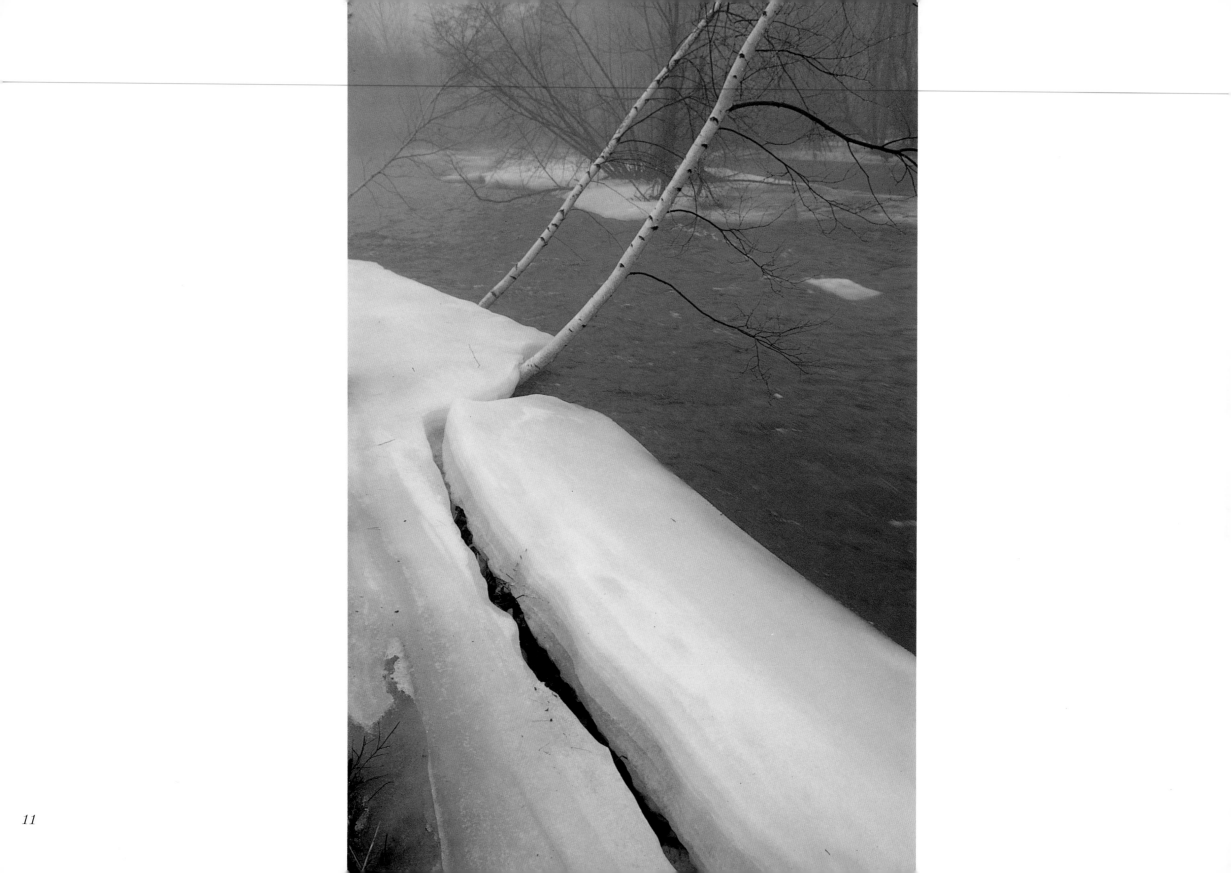

Then, perhaps in March, come a few warmer days and the silver tinge of melt appears on the edges of the banks, bright against black rock and water. Almost imperceptibly the water line grows. The ice that had choked and tightened the river is forced back, losing its terrain. There is a certain softening of the wind; gone is the frigid bite, the numbing blast. Along the shores of the tributaries, the waters, flecked with earth, drip steadily into the main flow.

This is the river's bloated time—a dirty, swollen bridge between winter and spring. Soon there is water in every gully, water dripping from tree and rock, water running in a thousand rivulets, filling every low place in the land, flushing out the old year's decay. The river is a wall of brown and slate water, and it will take every untethered thing in its way as it rushes downstream.

18

The air is warmer than the earth now, and as
soft rains fall the snowpack retreats and leaves
the ground swollen everywhere with water,
water that runs to the potholes, to the low points
near the bases of trees, building sloughs,
and backwaters, and bogs.

A misty warmth hovers over the river valleys.
From the droplets of water on the leaves to the
seepage from the hills, lush moisture is
everywhere.

Slowly, the vestiges of winter vanish. The water grows clearer. Along the banks a stripped fallen log shows where a beaver has been at work, and scattered among the trees are the first faint touches of spring.

Defying a late snow squall or a blast of frost, the world of the river is suddenly green and vitally alive. Air and earth suck up the excess water, gentler rains fall; the river begins to calm, and everywhere there is a warm expectancy.

Hellebore is often first, thrusting up out of the
moist earth, nourished by the mulch of decaying
leaves, its bright green plumage full against the
umber earth. The river water, clean and clear,
tumbles in a thousand falls. Everywhere there is
green, the bright, fresh green of spring.

24

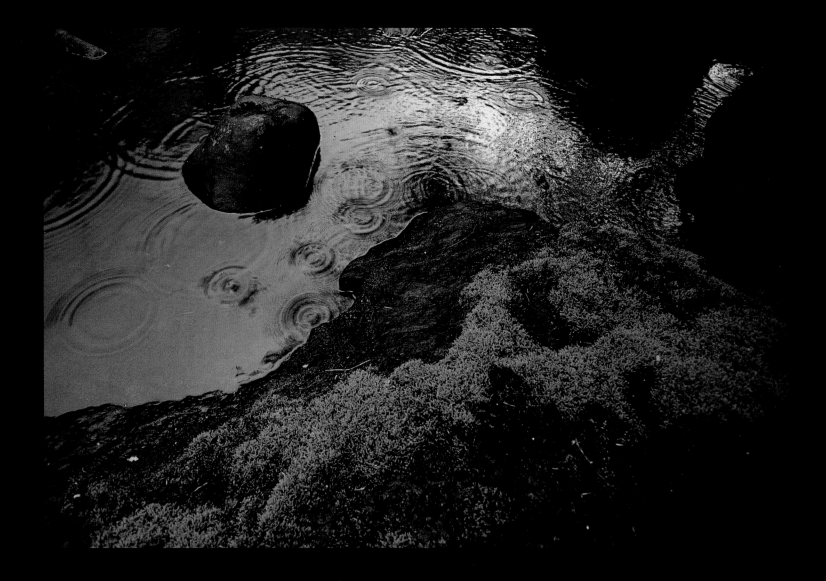

Shrubs, grasses, and mosses push their way through the moist earth. Leaves unfold. And from their buds spring a profusion of flowers: violets, yellow trillium, May apple, marsh marigold—an explosion of color.

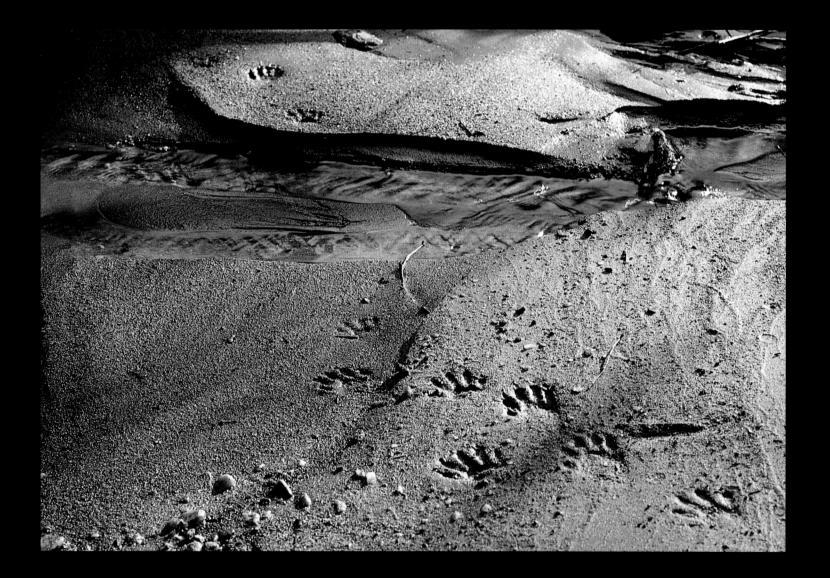

With the river as its locus, the animal world slowly begins to emerge from hibernation. A raccoon leaves tracks where he had prowled the riverbank looking for a frog, or a bruised minnow. A red squirrel near a budding oak peeps out of his hole to meet the new year; and there appear a bright red salamander, and a box turtle, and a toad. A woodcock, fresh from Louisiana, sweeps through the greening grasses, and from somewhere deep in the woods comes the heart-stopping sound of a grouse drumming to its mate.

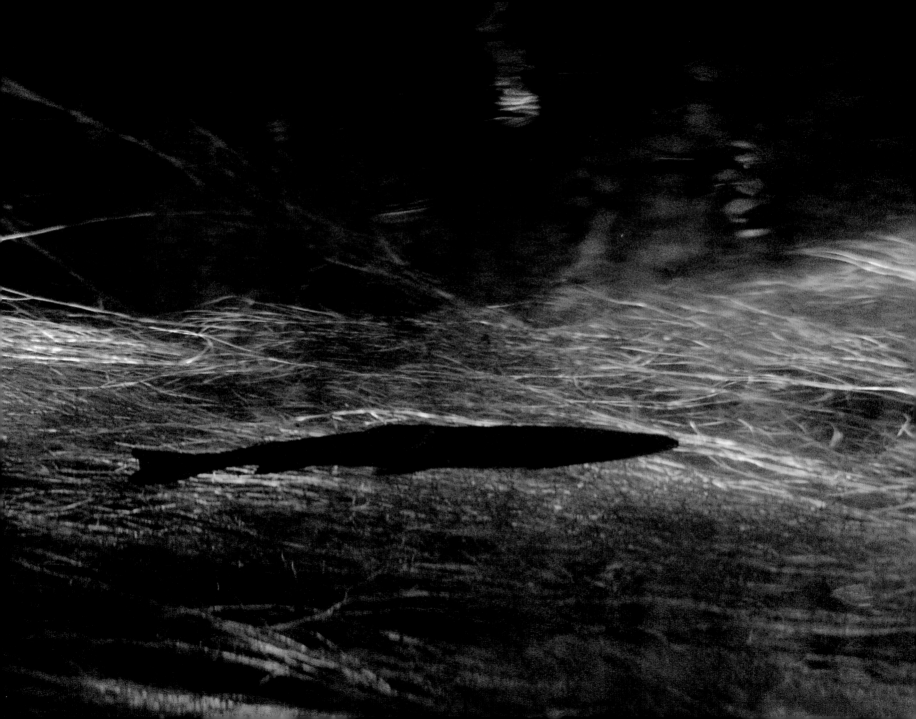

As the water warms, the first dark forms of trout slide out into a shaft of light. They rise closer to the surface now, visible, luxuriating in the sun, alert for food. They are a wonder to see: finning delicately over a weed bed, perhaps rising suddenly—with a gesture at once gentle and explosive—to take a fly from the roof of the river.

One warm afternoon the first mayfly begins tentatively to hatch, and a new cycle has begun. For the fly fisherman, mayflies are released from their Latin names in the entomology textbooks and named for the artificial patterns that, with cunning melds of fur and feather, imitate the natural insects. Now, in the wondrous litany of spring, they will all come in measured sequence: first, usually the Quill Gordon, then the Hendrickson, the March Brown, the Cahill, the Dun Variant, the Grey Fox, the Sulfur, the Green Drake.

The mayfly's life out of water lasts a day, perhaps several days, and then it is over. The entire crop of Hendricksons will hatch for a week, ten days, and a new species may start hatching before they are done; then another species starts, each in its own time, marking the progression of the spring as surely as a calendar. How gossamer they are—lightly veined, fluttering, like painted snowflakes in the wind. How delicate they are—the mayflies.

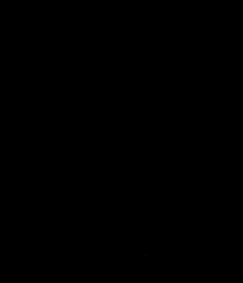

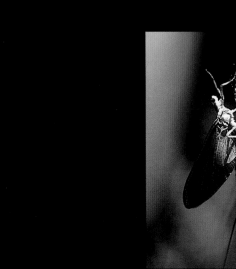

Now a new creature comes to the river, one who loves it, perhaps, more intensely and intimately and knowingly than any other visitor. He is the trout fisherman—and to him a river is forever intriguing. He will know it, not merely through surface impressions, but also out of an urgent need to understand its vital inner life. He must understand its currents and eddies—and where his quarry will be found, and when. He must know its multitude of stream life—the nymphs, mayflies, caddisflies, stoneflies, and minnows, local terrestrials that the trout will eat. Before his season is over, the fisherman will know the river at odd hours, when the rest of the world is asleep, before the mists of early morning are burned off by the bright heat of day, and after the dusk falls and the river is transformed once again into a dark world all its own.

Down in the meadows, the river runs clear and
low. This is the fly fisherman's finest hour.
He comes to be part of the wondrous bloom of
spring: he hears the frogs in the high grasses
and the drumming of the grouse; he sees
everywhere the exuberant yellow-green of spring.

The trout fisherman comes quietly, seeking
solitude, plying an ancient skill, and he will
leave the river the way he found it. He takes his
greatest pleasure in understanding the water he
fishes, in matching his artificial fly to the natural
insect.

If his choice has been correct, and if his cast
to a rising trout has been accurate, he may be
rewarded with that bulge on the surface that is a
trout taking a fly. Then the fisherman strikes, the
line grows taut. He is suddenly connected
to something live and wild.

In its good time, the fish will tire and may be led to net.

There is a special beauty to the fish as it comes close, gorgeously mottled against the haunting riffles and angles of the water. Then there is a moment for appreciating the pied beauty of the trout—its sleek form and glistening, spotted back—a moment of final possession.

As the season changes from spring to high spring to early summer, the fisherman feels the need for broader experience. He will begin to venture out to waters farther from home. Wherever there is free-flowing clear water you will find him. He will be on the great freestone rivers of the West that plummet through miles of granite mountain passes and on the mighty steelhead rivers of the Pacific Northwest and Alaska. He will come to the delicate meadow creeks of the East when he thinks the river is at its optimum—or when he can come.

According to his temperament, he will prefer
hard-tumbling water, pocket water, flat glides,
spring creeks, or a high meadow river with its
pools, bends, runs, and riffles.

Such rivers are scattered from North Carolina to Maine, from New England to California, in states such as Arkansas and Arizona, with great clusters in Montana, Idaho, Michigan, Pennsylvania, New York, New Hampshire, and Vermont. Each river, stream, or brook is given its own wiggly line on a map and each has a name.

Spruce Creek,

a limestone in Pennsylvania

The heavy, broken rush
of the lower Madison in Montana

The broad, sweeping shallows
of the historic Susquehanna

The weedy, trouty, food-rich waters
of Armstrong Spring Creek in Montana

The Whitewater, a happy little mountain creek
in the Smokies of North Carolina

Raven's Fork, a tumbling creek
in the Blue Ridge Mountains

The rock-bound pocket water
of New York's Ausable River

To the fisherman these are names rich with association. Some are derived from Indian or Dutch origins, some commemorate a moment in history, a President, an explorer, a place, a local event, a geographic feature. How different, how keenly their own, are each of them.

BEAVERKILL RIVER, NEW YORK

DESCHUTES RIVER, OREGON

FRYING PAN RIVER, COLORADO

GARDINER RIVER, WYOMING

LAMOILLE RIVER, VERMONT

MARGAREE RIVER, NOVA SCOTIA

DELAWARE RIVER, NEW YORK

WOLF RIVER, WISCONSIN

BIG HOLE RIVER, MONTANA

LETORT RIVER, PENNSYLVANIA

ROGUE RIVER, OREGON

SHENANDOAH, VIRGINIA

FIREHOLE RIVER, YELLOWSTONE NATIONAL PARK

AU SABLE RIVER, MICHIGAN

83

KLAMATH RIVER, CALIFORNIA

NORTH UMPQUA RIVER, OREGON

GUNNISON RIVER, COLORADO

CHATTOOGA RIVER, SOUTH CAROLINA

GOOD NEWS RIVER, ALASKA

For the fisherman summer comes too soon.
In late June, July, and into August, the water
level of the river drops. Under the hot summer
sun, reeds and water lilies luxuriate in the
backwaters. Everything is lazy, sleepy, slow.

Fewer fishermen come to the river in high
summer, and when they do they find their
favorite stretches of water shrunken,
transformed, often filled with swimmers or
canoers. Most of the mayflies are gone and the
trout feed chiefly during the extremities of the
day, in early morning or at dusk. Often,
to escape the increasing heat of the water,
the trout stack up by the dozens in the mouths of
cooler feeder creeks.

The fisherman who cannot leave the river he loves may fish more slowly now, with less expectation, less fever for his sport. He may kick a stone or two, study what two months earlier was the opaque interior of his river.

He may examine stonefly cases on a rock, or one left when a stonefly crawling from the water to hatch, looking for all the world like some miniature prehistoric monster, climbed a stick and there shucked its case.

He may see a spider's web where once there had been water, or an open mussel shell, left perhaps by that raccoon whose tracks sign the soft, wet sand.

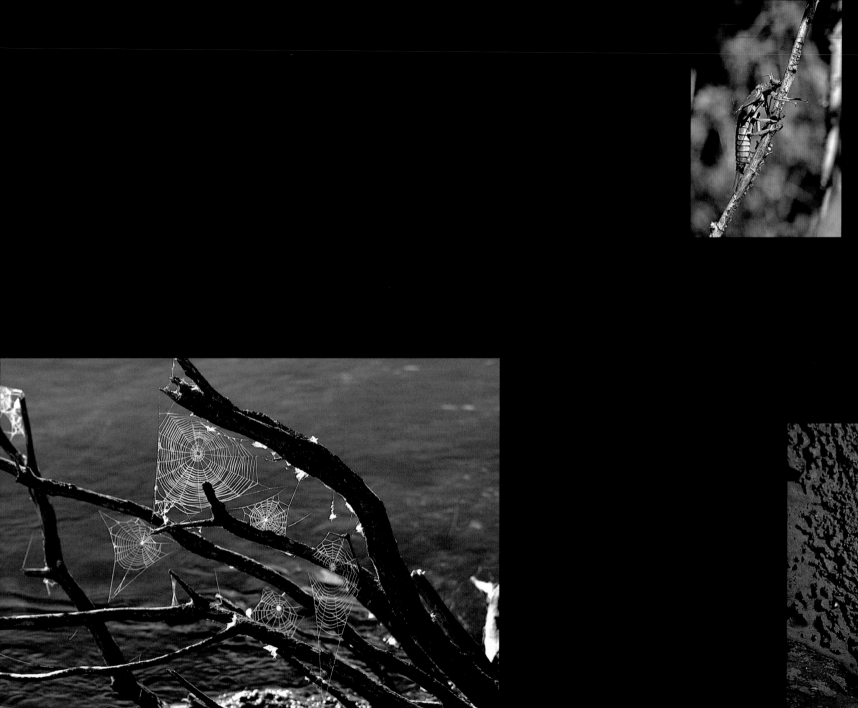

Here, too, a sandpiper, pecking for insects where the river has retreated, has left its tiny footprints. And then, a lazy trout plucks a tiny terrestrial from the surface of the water with a rise so slow you might have thought it had only, from boredom, bumped the roof of its world.

In the fields are grasshoppers, butterflies, crickets, and the trout lie patiently in wait for them to fall or be blown into the water by an afternoon gust of wind.

In summer the world of the fisherman's river is still and dull and peaceful.

Wherever he is, the fisherman finds the coolest part of the day best. He will keep mostly to the fringes of the day and will often find the evening and night most productive. As the shadows of the evening grow longer, the river becomes quite different again. Distances blur; sounds magnify; a sweet silence—broken only by the songs of frogs, crickets, and birds—blankets everything.

Now man becomes one with the mule deer,
the great horned owl, and the other creatures of
the night. The fisherman, tired from the heat of
summer, slips into the dark stream, welcomes the
refreshing cool of air and water, and alone, casts
toward the susurrant sounds of feeding fish.

By September the year has ripened and the great thrust of life in gourd, leaf, and fruit has reached its peak. Autumn will be a mix of fullness and melancholy, of chill dawns and dusks and bright days. John Keats called it the "season of mists and mellow fruitfulness." The year will press all living things to their optimum richness and then, a moment later, each living thing will begin to die.

Down in the valley, the cooler air of approaching winter meets the warm river. In the mornings, coming to the river through the long and lonely fields, everything seems golden: grasses everywhere are tinged with gold, and the sun—through the oaks along the river—takes on a Midas glow.

Seen in its smallest instance or in panorama,
autumn comes in now with a blast of color.
The gold is soon mixed with crimson, umber,
russet, yellow, bronze, rose—every warm hue
in the artist's palette.

In autumn the fisherman returns. Mostly he has come to be near the river he loves. He wants to see and to remember every feature of it, every rock and riffle and eddy, during the long winter ahead. He sees where beavers are shoring up their dams and getting ready their food supply. He flushes a woodcock that had been resting overnight in its flight from New Brunswick to the Deep South.

There are trout to be caught and he will take his share, for the fish are hungry, storing up for the winter, less selective in their feeding now. From a pool he takes several brookies—bright crimson and black—and lays them on a stump in the river to measure their color against the color of the day.

124

The fisherman has returned to see and to hold,
and what he sees most in autumn, as the season
grows older, is color. The infinite variety begs the
viewer to stop, insists he wonder at the million
details, brilliantly composed, on and near the
water. Nature spills over, explodes.

See how the old, dead elm tree slants at such a pleasing angle, how its form gives way to the ghostly rocks beneath the surface, while its bleached trunk and crimson and orange leaves bounce their reflections against the cobalt water. A soft breeze blows and, moments later, a sharper, gustier puff of wind breaks these patterns into pure abstractions of soft-edged shapes in lovely pastel shades.

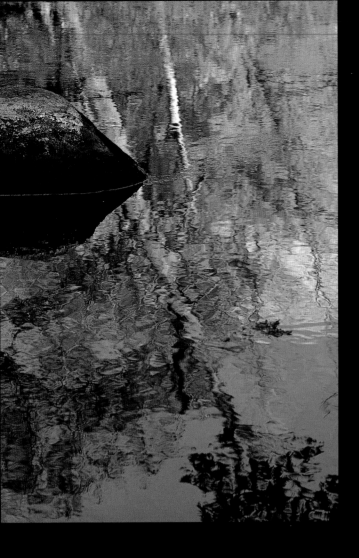

Caught in such absorbing scenes, the fisherman
may not feel at first the telltale chill, the
harbinger of winter. He may not notice how
much of the world has grown umber, dark russet.
But one day as he shoots his fly line toward the
left corner of a midstream boulder, a place where
he knows a good trout waits, the line drops
short. It has lost its customary grace, its fluidity.
Ice has formed at his guides and his breath,
like a little cloud, has turned to mist.

Now the signs are everywhere. An icicle hangs from the branch of an oak sapling. In a backwater, skim ice has trapped the fallen leaves that had been riding the currents so gaily.

There are a thousand other hints of winter, a thousand exquisite icons of late fall.

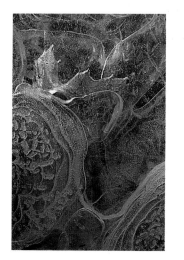

High in the mountains, storm clouds gather steadily; some have already dusted the whole range with snow. Winds gust most of the day now, and the long, stark branches of the trees shake and shiver from their force. The air has a sharp bite to it and slaps cheeks into a bright red. Winter is near. Each day more of the faint remaining color leaves the landscape; each day there are fewer birds, fewer sounds, more everywhere that is brushed with frost and early snow.

142

There is a bittersweet quality to these last lingering days of the fisherman's year. He is reluctant to leave, reluctant to let go of the season. These are the days he wants to hold back, snaring one or two more from the inescapable thrust of cold—melancholic days, touched with feeling.

And then there is a day when the fisherman
decides that, against all his longings, he must
take down his rod, put his reels in chamois
bags, tuck all his flies into their proper boxes,
and store everything, along with his memories,
for the winter. He will fish no more until spring.
His season is over.

146

Now the whiteness grows more rapidly. First
it laces only the tops of the branches; then it
lightly covers the cold earth between the trees.
It grows wetter, heavier. It plasters the windward
trunks of trees in cold symmetry, leaving less
and less, anywhere, that is black. It heaps and
piles and rounds itself on the tops of rocks and
hillocks, softening the edges of the landscape.
It covers the midstream boulders and builds with
the ice along the rim of the river.

For a time, little will grow. Increasingly, the survivors of the past year will fly to warmer climates, hibernate, or die. Soon the full force of the cold will descend. And the river, once the source and fulcrum of life, is left to itself again.

But to the true lover of rivers this actually may
be the time of year when the stream is most
uniquely beautiful. The lone wanderer along its
banks is dazzled by the silvered sheen,
the omnipresence of white, the brightness,
the repetition of forms.

Held in wonder by such whiteness and
harshness, he scans it, walks through it gingerly,
studies it for the special hieroglyphics, for some
sign of life. He finds it across a windswept pond,
the winter home for a beaver. He startles a
grouse, and when he turns a bend in the river he
spots a pair of deer watching him from the far
side where they had been browsing. As he looks
up he sees the beady eye of a red squirrel
peering at him from behind the limb of a beech
tree. Life seems as sparse and fugitive as that
chickadee huddled on a bare branch.

160

It is deep winter again and the river is quiet and cold. There is a mystery to the landscape now. For months only a spare and solitary life will be lived along the river's shores and in its depths.

In the silent landscape little dramas are played out. Along the stream bank are the tracks of a red fox. It has been hunting, perhaps for some unsuspecting mouse. The tracks—alone breaking the smooth softness of the snow—lead into the hills. They leave the river, wrapped in its haunting white cloak, regaining itself, as vital as ever, running steadily on, toward its appointment with another spring.

Editor: Robert Morton
Designer: Elissa Ichiyasu

Library of Congress Cataloging-in-Publication Data
Madison, Lawrence.
Trout River.
1. Trout fishing.   2. Stream ecology.
I. Lyons, Nick.   II. Title.
SH687.M293   1988   799.1′755   87–38509
ISBN 0–8109–1697–5
Published in 1988 by Harry N. Abrams, Incorporated, New York.
A Times Mirror Company
Printed and bound in Japan